Atlas of
British Columbia
People, environment,
and resource use

ON JUNE 25TH A NEW ATLAS OF BRITISH COLUMBIA WILL BE PUBLISHED.

ENCLOSED IS PERTINENT INFORMATION DESCRIBING THE ATLAS AND THE AUTHOR.

WE HOPE THAT YOU WILL FIND SOME OF THIS BACKGROUND MATERIAL USEFUL.

The University of British Columbia Press
2075 WESBROOK MALL, VANCOUVER, BRITISH COLUMBIA, CANADA, V6T 1W5
TELEPHONE (604) 228-3259 / CABLE: UNIPRESS VANCOUVER

The University of
British Columbia
Press

New Books
Spring

1979

The University of British Columbia Press

2075 Wesbrook Mall, Vancouver,
British Columbia, Canada V6T 1W5
Telephone: (604) 228-3259
Cable: Unipress/Vancouver

The University of British Columbia Press continues to concentrate its scholarly publishing in the following areas: *International Law, Asia and the Pacific,* and *Canadian Studies*. Books of special interest outside these areas are published from time to time.

In addition to Spring 1979 books this booklist features titles of continuing interest.

Cover illustrations: from ATLAS OF BRITISH COLUMBIA

Printed in Canada

Table of Contents

Canadian Cataloguing in Publication Data
University of British Columbia Press books contain CIP Data

International Standard Book Number
Publisher's ISBN prefix is 0-7748-

New Books Spring

1979

Atlas of British Columbia

People, Environment, and Resource Use

A.L. Farley

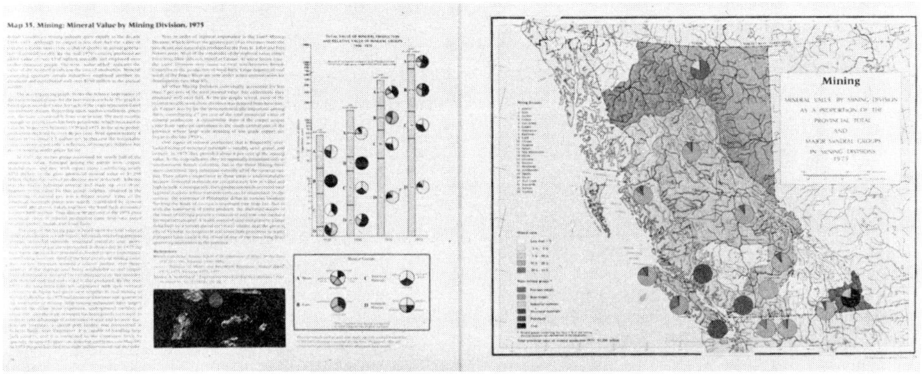

The *Atlas of British Columbia* is the first major cartographic study of the province to be published since 1956. Created through close co-operation between government, the private sector, and the university, it is the successor to the British Columbia Atlas of Resources which, for twenty years, has been the standard reference work used by schools, industry, government, and the general public.

The most recent data available have been used to give an accurate, comprehensive picture of British Columbia's economy as it is today. Comparative studies show the development of the province's manpower and natural resources as well as the rapid growth of industry and technology since the beginning of the century.

For easy reference, the three main subject areas are divided into 61 topics — each illustrated by a map page and a facing page of interpretive text. They include:

THE PEOPLE: changes in settlement patterns; distribution of native peoples; growth of the labour force.

THE ENVIRONMENT: geological, topographical, and climatic features; concentration of fish and game.

RESOURCE USE: distribution of natural resources; development of industry; communication and transportation links; location of energy resources; development of recreational lands.

The atlas has 115 maps, 74 illustrations, tables and diagrams. Bold colours and explicit keys contribute to the easy, rapid interpretation of data. A *Gazetteer* and an *Historical and Statistical Summary* are included to provide a quick reference for the most commonly sought facts and to supplement maps and text.

The atlas is an indispensable reference tool for government officials, businessmen, investors, industrialists, teachers, students, librarians, environmentalists, and all people with a fundamental interest in the past, present, and future development of British Columbia.

A.L. FARLEY was a cartographic editor of the British Columbia Atlas of Resources (1956) and is on the faculty of the Department of Geography at the University of British Columbia.

144 pages, 28 x 32.5 cm (11 x 14 in.), May 1979
illustrated, 33 colour, 41 black and white photographs
11 tables and 16 diagrams
ISBN 0-7748-0092-5
cloth, **$45.00 Tr.**

Table of Contents

SECTION A — PEOPLE 24 maps, 22 pages

POPULATION: Historical — 1911-1931; -1976; NATIVE INDIANS: Distribution - 1850; Occupied Reserves and Proportion of Status Indian Population in Total Regional Population, 1974; EXPLORATION; THE "TRUTCH MAP" 1871; LANDSAT MOSAIC, 1975; MIGRATION, 1966-1971; NET MIGRATION & CHANGE, 1966-1971; MIGRATION, 1971-1976; LABOUR FORCE, 1951-1971.

SECTION B — ENVIRONMENT 27 maps, 30 pages

RELIEF AND BATHYMETRY; BEDROCK GEOLOGY; PHYSIOGRAPHIC REGIONS; VULCANISM, SEISMIC ACTIVITY, AND MAJOR GROUNDWATER FEATURES; GLACIAL FEATURES; SOILS; STREAMFLOWS; WATER QUALITY; MEAN ANNUAL PRECIPITATION; SEASONAL TEMPERATURE AND PRECIPITATION; FROST FREE PERIOD AND MOISTURE INDEX; BIOGEOCLIMATIC ZONES; LAND CAPABILITY FOR WILDLIFE-UNGULATES; GAME ANIMALS: Distribution and Regional Harvest, 1974; FRESHWATER SPORTFISH: Distribution and Catch, 1969-1970.

SECTION C — RESOURCE USE 64 maps, 70 pages

FORESTRY: Distribution of Forested Land; Distribution of Commercial Species; Forest Administration, 1976; Location and Capacity of Sawmills, — 1931-1961; -1971; Pulp and Paper Mills, 1911-1961; Pulp and Paper Mills and Plywood Plants, 1975; MINING: Historical — Value of Production by Mining Division, 1909-1973; Mineral Value as a Proportion of the Provincial Total and Major Mineral Groups, 1975; Mined Minerals, Producing and Potential Mines, and Smelters, 1976; AGRICULTURE: Historical, 1911-1971; Agricultural Land Reserves and Livestock Distribution, 1975; Capitalization as a Percentage of the Provincial Total and Farm Types by Major Enterprise, 1971; Crop Value by Census Division as a Percentage of the Provincial Total and Crop Types in Census Division, 1971; Livestock Value by Census Division as a Percentage of the Provincial Total and Livestock Enterprises in Census Division, 1971; FISHING: Historical — Distribution of Operating Fish Processing Plants and Salmon Canneries, 1911-1975; Average, 1971-1975; Salmon Net and Troll Fishing and Major Salmon Spawning Streams; ENERGY: Historical Development of Electricity Generation — 1911-1971; Electricity Generation and Transmission, 1975; Potential Generation Capacities, Fossil Fuels, and Pipelines; Electrical Energy Consumption by Electrical Service Areas, 1961 and 1971; WATER USE, 1976; MANUFACTURING: 1951, 1970; -1972; RECREATION AND TOURISM: Provincial and National Parks and Marine Parks, 1975; Ski Areas and Boat Launching Ramps, 1975; Total Campground Attendance, 1975; Campground Attendance and Summer Vehicular Traffic Volumes, 1975; Flow of Campers to Selected Regions, 1976; TRANSPORTATION: Land and Ferry Routes and Airports, 1976; Traffic Flows (Percentage Change 1966-1971); Rail Flows by Major Commodity Group (Mean Sample Values 1972-1974); COMMUNICATION: Regional Linkages Based on Telephone Calls, 1975; GAZETTEER MAP, HISTORICAL AND STATISTICAL SUMMARY.

God's Galloping Girl

The Peace River Diaries of Monica Storrs, 1929-1931

Edited by W.L. Morton with the assistance of Vera Fast
Preface by R.D. Symons

What brought Monica Storrs to embark on a wilderness life in the depressed thirties amidst the hardships of B.C.'s Peace River country — the last North American frontier?

Monica Storrs was to stay in the north for twenty years. She came to a wilderness almost roadless, townless, and schoolless, without amenities or culture; a land of hot, dusty summers and unpredictable, sub-Arctic winters; a land of untamed and unexpected grandeur.

Hearing the "clear call" of duty, she was endowed with a strong desire to serve the Anglican cause. A refined woman of 41, the delicately reared daughter of the dean of Rochester Cathedral, she was following a family tradition in seeking missionary work in rough and alien surroundings.

Her mission was the settlers and the frontier. She came to teach Sunday School and the traditions of guiding and scouting to tough, work-hardened country children. She rode horseback to bring the gospel to the families of isolated farmers and trappers. She welcomed the poor and hungry to her home — both the god-fearing and the godless. And she was, for the most part, welcome in theirs.

She and the women who became her helpers were known affectionately by the people of the Peace as "God's Galloping Girls." Overcoming early fears of failure and rejection, she learned to subdue the prejudices of her moral and social upbringing and at last won acceptance and a permanent place in the heart of the north.

Monica Storrs' diaries are a moving narrative of a woman's indomitable courage, faith, and humour. Her story is a vivid, realistic description of frontier life and the struggle for survival. Her words convey the wild and primitive beauty of the Peace River country — not just a region, but an adventure of the spirit.

W.L. MORTON is professor of history at the University of Manitoba. He is the author of many books and articles on Canadian history.

400 pages approx., 15 x 23 cm, August 1979
illustrated
ISBN 0-7748-0110-7
cloth, **$22.00 Tr.**

God's Galloping Girl is the third volume in a series of editions of important documents of the colonial and early provincial history of British Columbia. Other books in this series are:

A Pioneer Gentlewoman edited by Margaret A. Ormsby
cloth, **$18.95 Tr.**, paper, **$7.95 Tr.**

The Reminiscences of Doctor John Sebastian Helmcken edited by Dorothy Blakey Smith, cloth, **$18.95 Tr.**

Harsh and Lovely Land

The Major Canadian Poets and the Making of a Canadian Tradition

Tom Marshall

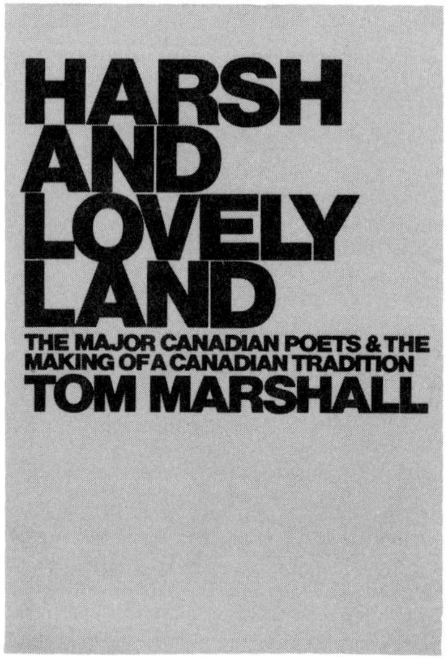

"My central concern has been the poem that is Canada and Canadians. . . . The journey of the Canadian poet. . . ."

Poet-critic Tom Marshall examines four stages in the development of a purely Canadian tradition in poetry through a focus on the work of major poets writing in English from the mid-19th century to the present.

While recognizing the danger of labelling any characteristic as peculiarly "Canadian," he views certain recurrent themes as Canadian in their formal development.

He begins with the "Pioneers" such as the "Confederation" poets Lampman, Carman, and Roberts, who had the task of adapting British classical traditions to the raw experience of the new land and who set the stage for D.C. Scott and E.J. Pratt. The "Modernists" who followed brought a new sophistication to Canadian poetry and made possible the combination of vigour and subtlety found in the best work of poets like Earle Birney, Irving Layton, and A.M. Klein.

The third phase, the "Inheritors," involves the emergence in the work of such poets as Margaret Avison and Al Purdy of a distinctively Canadian "voice" and idiom. Lastly, Marshall focuses on the work of the "Poet-Novelists" Margaret Atwood, Leonard Cohen, and others. This recent literary development he describes as a purely Canadian phenomenon.

Full of sharp insight and infused with poetic imagination, this book is a valuable contribution to the study of literature in Canada.

TOM MARSHALL is an associate professor of English at Queen's University. He is former editor of *Quarry* and has published several books of poetry and literary criticism.

200 pages approx., 15 x 23 cm, May 1979
ISBN 0-7748-0107-7
cloth, **$18.50 Tr.**

Planning/Urban Studies

The Challenge of Squatter Settlements

With Special Reference to the Cities of Latin America

Harry A. Anthony

Twenty-three to fifty per cent of the urban population of the underdeveloped countries of the world are squatters whose cheaply constructed shacks are located in the squalid areas of cities.

Harry Anthony draws dramatic attention to the plight of these migrant poor. He restricts his study to the squatter settlements of Latin America where the issues are more complex, and solutions more urgent, and where cities are growing at a rate unprecedented in human history.

While recognizing the colossal economic, social, and political problems posed by these settlements, the author sees them as vital communities populated by resourceful people who could make a positive contribution to the progress of their countries.

He cites the failure of such programmes as forcible relocation to deal with the crisis and describes and evaluates policies and projects which are already achieving positive results. He concludes his study with the official United Nations Habitat recommendations which have direct bearing on the subject.

In researching this human settlements issue, the author studied hundreds of reports and videotapes presented at the Habitat '76 Conference. He has also drawn on his experiences as a planning consultant in many of the world's developing countries.

HARRY A. ANTHONY is professor of urban planning at the California State Polytechnic University. He has published many articles and reports on urban planning and development issues.

90 pages approx., 15 x 23 cm, April 1979
illustrated
ISBN 0-7748-0111-5
paper, **$4.95 Tr.**

The Challenge of Squatter Settlements is the third in a series of studies published in association with the Centre for Human Settlements. Other books in this series are:

Habitat and Land by Len Gertler, **$4.95 tr.**
Planning and Building Down Under by Harry Seidler, **$4.95 Tr.**

Anthropology/Sociology

Malaysian Mosaic

Perspectives from a Poly-Ethnic Society

Judith Nagata

The Malaysian case is a prime example of the difficult and often conflict-ridden task of nation-building in a "plural society." *Malaysian Mosaic* documents the growth of ethnic communities in Malaysia and examines how the identity of ethnic groups is retained in a society undergoing radical modernization.

This book combines traditional anthropological material from small communities with a more encompassing overview of the impact of ethnicity at the national level. It attempts to show how far ethnic identities are affected by social, economic, and political interests and how ethnic roles interact with those of class, caste, "status group," and kinship.

The historical background describes the origin of ethnic communities in the early, frontier-like Malay Peninsula and examines the conditions under which ethnic differences develop. Lesser-known groups from Siam and Indonesia, in addition to the main ethnic "trinity" of Malays, Chinese, and Indians, are discussed in detail.

The problems in contemporary Malaysia are most evident in the urban centres. The author describes the traditions and ranking systems of the major ethnic groups and the extent to which they are becoming "submerged" under modern influences. Field studies of life in two different types of towns highlight the conditions of urban life which develop and sustain ethnic identities.

In concluding, the author incorporates this material in an intensive and systematic analysis of the nature of ethnic identity and challenges many of the assumptions and stereotypes of the Malaysian legal system as well as present social change policies.

This book will be of interest to sociologists, anthropologists, social historians, and all those interested in the problems of integration in a multi-ethnic society.

JUDITH NAGATA is associate professor and director of the graduate programme in social anthropology at York University.

288 pages approx., 14 x 21.5 cm, June 1979
1 map
ISBN 0-7748-0106-9
cloth, T.B.A.

Labour Relations/Japan

The Japanese Working Man

What Choice? What Reward?

Ernest van Helvoort

From the shambles of the Second World War Japan emerged to become an industrial giant in the free world. Why is Japan so successful? How has she managed to secure such a firm hold on the international marketplace? This increasingly important question preoccupies Japan's major trading partners, who are currently having to come to terms with her immense balance of payments surplus.

There are many reasons advanced to rationalize Japan's economic and industrial achievements, but in the view of Ernest van Helvoort, Japan's success lies to a far greater extent than is appreciated in what he calls the *human factor* as it is demonstrated by Japan's remarkable personnel management system.

How are people motivated to render their services? What are the basic principles of reward and promotion? How are training and education conducted in relation to industrial needs? What are the particular Japanese features of participation, communication, and decision-making? Is the average Japanese worker happy in what he is doing and content with his lot in life?

These and many more fundamental questions are probed in depth by the author. In the final analysis he points out that, however unwholesome and discriminating certain aspects of the employment structure may be, the Japanese working man is part of a culture that has learned to adjust with the changing times; it has learned to turn misfortune from without or shortcomings from within to its own advantage.

ERNEST van HELVOORT is author of the study *Blue-Collar Workers in Japan and Holland* (1977). He is director of the Netherlands National Tourist Office in Tokyo.

150 pages, 15 x 23 cm, April 1979
ISBN 0-7748-0109-3
cloth, about **$16.00**
FOR SALE IN NORTH AMERICA ONLY

International Law

The Canadian Yearbook of International Law Annuaire Canadien de Droit International

Volume/Tome XVI, 1978

Edited by C.B. Bourne

Annually since 1963 the *Canadian Yearbook of International Law* has presented an impressive and authoritative survey of important and topical issues in the field of international law, particularly where these are relevant to Canada.

In addition to articles of lasting significance in the field, each volume contains book reviews, a Notes and Comments section, two sections on Canadian practice in international law, and a digest of important Canadian cases on public international law and conflict of laws.

Articles in Volume XVI include: "The Scottish Tradition in International Law," "Les accords bilatéraux: régissant la pêche étrangère dans les eaux canadiennes," "Derogation of Human Rights in Emergency Situations," "Des normes régionales et universelles pour les droits de la personne," "The Status of Agents on Special Mission in Customary International Law," "The Legal Status of Agreements Concluded by Component Units of Federal States with Foreign Entities."

Volume XVI continues the high standards described by Professor Wilhelm Karl Geck in his review of earlier volumes in the series: "Canada's growing international importance and its increasing interest in international law with special emphasis on problems of the sea, will make the *Yearbook* particularly attractive to international lawyers the world over...." *The Canadian Yearbook* is issued under the auspices of the Canadian Branch of the International Law Association (Canadian Society of International Law).

C.B. BOURNE is professor of law at the University of British Columbia.

440 pages approx., 16 x 23.5 cm, June 1979
ISBN 0-7748-0112-3
cloth, **$22.00**
A limited number of Volumes I to XV is available.
A complete set — Volumes I to XVI will be sent for $254.00 pre-paid.

Contact and Conflict

Indian-European Relations in British Columbia, 1774-1890

Robin Fisher

NOW IN PAPERBACK

Winner of the 1977 Sir John A. Macdonald Prize for the non-fiction work of Canadian history "judged to have made the most significant contribution to an understanding of the Canadian past."

"a well-written, stimulating book recommended for university and college libraries with an interest in Canadian and Indian history."

CHOICE

"without doubt one of the best accounts that has been written. It is a book which will be in demand by every student of British Columbia history and certainly should be on the bookshelf of those doing anthropological studies in the area."

AMERICAN WEST

"his research opens new insights for both historians and anthropologists."

CANADIAN ETHNIC STUDIES

In this thought-provoking book, Robin Fisher traces the role of the Indians in the history of British Columbia and shows how the reciprocal nature of Indian-European relations in the early days of the fur trade became a pattern of white dominance.

ROBIN FISHER is a member of the history department, Simon Fraser University.

268 pages, 15 x 23 cm, February 1979
illustrated, map
ISBN 0-7748-0108-5
paper, **$8.95**

(also available
268 pages, 15 x 23 cm, 1977
illustrated, map
ISBN 0-7748-0065-8
cloth, **$18.00)**

One Century Later

Western Canadian Reserve Indians Since Treaty 7

Edited by Ian A.L. Getty and Donald B. Smith

PREVIOUSLY ANNOUNCED

1977 was the centennial of the signing of Treaty 7 — the final treaty in which the native peoples of Southern Alberta surrendered their land to the Crown. Prominent native and non-native authorities contribute thought-provoking commentaries on the diverse experiences of Western Canadian Indians before and after 1877.

Details of life on the reservations in the post-treaty period and government attitudes and policies in implementing the treaties are examined — subjects which, until recently, have been largely ignored by historians.

A new perspective offers fresh insight into the role of the Indian in the fur trade, and case studies chronicle the history of the Sioux in Canada and the life of the Plains Indians in the 1920's and 1930's. Contemporary issues — taxation, education, housing, unemployment, social services, natural resource development — are examined in relation to treaty provisions.

Many current national and provincial arguments involving Indian rights have their roots in the frustrations and misunderstandings arising from the treaties. *One Century Later* is a dynamic voice that speaks directly of these experiences.

The papers in this book were originally presented at the Western Canadian Studies Conference in 1977.

IAN A.L. GETTY is a doctoral candidate in history at McMaster University.

DONALD B. SMITH is associate professor of history at the University of Calgary.

190 pages, 15 x 23 cm, February 1979
illustrated, tables
ISBN 0-7748-0103-4
cloth, **$16.50 Tr.**
ISBN 0-7748-0105-0
paper, **$6.95 Tr.**

Transport in Western Siberia

Tsarist and Soviet Development

Robert N. North

PREVIOUSLY ANNOUNCED

"The expansion of Russian activities east of the Urals is one of the most fascinating examples of large-scale settlement and economic development ever recorded...few regions have more to offer the student of cases."

This highly specialized work is the most comprehensive study in English of the economic history of the western Siberian region.

Tracing the economic development of western Siberia and northern Kazakhstan from the 1580's to the present, the author focuses on one aspect of development in particular — spatial relationships as reflected in the volume, direction, and composition of freight traffic flows. He analyses these changing spatial relationships in nine sequential periods in terms of transport modes and freight traffic. The impact of political, economic, social, demographic, and technological factors is also examined within each period.

The extent to which the region might have developed differently under a less centralized, free-enterprise economy is discussed. Models of newly settled regions suggested by North American scholars are reviewed and their usefulness in explaining regional development in a different society is assessed.

This book is an important contribution to literature on the impact of the Soviet system and on the significance of the location of political-economic power in regional growth. It provides valuable insights into the general topic of economic development of continental interiors, and many of its theories will be useful to students of comparable North American regions.

ROBERT N. NORTH is associate professor of geography at the University of British Columbia.

448 pages approx., 15 x 23 cm, May 1979
maps and tables
ISBN 0-7748-0097-6
cloth, **$22.00**

Trade Titles of Continuing Interest

1979

One Century Later. Getty, Ian A.L., and Donald B. Smith (Editors). cloth, $16.50 Tr., paper, $6.95 Tr. Leading native and non-native authorities provide thought-provoking commentaries on the experiences of Canada's western native peoples since 1877 when the Indians of Southern Alberta surrendered their land to the Crown.

Ocean of Destiny. Lower, J. Arthur. cloth, $16.50 Tr. The first major study correlating historical events of over five centuries which took place concurrently on Asiatic and North American Pacific coasts.

Canadians on the Nile, 1882-1898. MacLaren, Roy. cloth, $14.95 Tr. Canadian youths joined England's young bloods in pursuit of "Gold, God, and Glory" on imperial exploits in the Sudan.

Exploring Vancouver 2. Kalman, Harold, Photographs, John Roaf. paper, $7.95 Tr. A popular architectural history guide to Vancouver's downtown landmarks and suburban districts of interest.

Early Indian Village Churches. Veillette, John, and Gary White. paper, $15.00 Tr. Winner of the American University Presses' award for design excellence, this unique photographic record preserves the image of B.C.'s wooden frontier architecture.

A Pioneer Gentlewoman in British Columbia. Ormsby, Margaret A. (Editor). cloth, $18.95 Tr., paper $7.95 Tr. Susan Allison's recollections contain all the excitement of living in pioneer times in B.C.'s interior wilderness between 1860 and 1890.

The Reminiscences of Doctor John Sebastian Helmcken. Blakey Smith, Dorothy (Editor). cloth, $18.95 Tr. A practising doctor with the Hudson Bay Co. in Ft. Victoria, Helmcken entered colonial politics and helped negotiate B.C.'s entry into Confederation.

Western Canada Since 1870. Artibise, Alan F.J. paper, $6.95 Tr. The first up-to-date, one volume guide to the most significant literature in the humanities and social sciences on all four western provinces.

For Most Conspicuous Bravery. Roy, Reginald H. cloth, $15.95 Tr. An account of Major-General George Pearkes' exciting career from Alberta homesteader to lieutenant-governor of B.C.

The Wind Commands. Morton, Harry. cloth, $29.00 Tr. A panoramic survey of Pacific sailing ships and of the great navigators, explorers and sailors who manned them.

Visitors Who Never Left. Harris, Kenneth B., in collaboration with Frances M.P. Robinson. paper, $5.95 Tr. The first English translation of important Indian legends compiled by a member of the culture they spring from.

Indian Masterpieces from the Walter and Marianne Koerner Collection in the Museum of Anthropology. paper, $4.95 Tr.

Northwest Coast Indians Artifacts from the H.R. MacMillan Collections in the Museum of Anthropology. paper, $4.95 Tr.

These catalogues present a colourful visual and general survey of two collections, which together, cover all seven native groups that inhabited the northwest coast area.

A Political Art. New, William H. (Editor). cloth, $15.00 Tr. Leading writers and artists on both sides of the Atlantic contributed original work to this salute to George Woodcock, Canada's foremost man of letters.

Tommy Brayshaw. Read, Stanley (Editor). cloth, $32.00 Tr. This collector's item features 29 of Brayshaw's famous salmon paintings and carvings in full-colour and pen and ink drawings, along with important biographical material.

The Royal Navy and the Northwest Coast of North America 1810-1914. Gough, Barry. cloth, $14.00 Tr. This book examines the hitherto unappreciated role of British warships in determining Canada's western boundary.

Planning and Building Down Under. Seidler, Harry. paper, $4.95 Tr. An internationally renowned architect examines new urban and regional planning concepts in Australia and discusses influences on contemporary architecture.

Habitat and Land. Gertler, Len. paper, $4.95 Tr. A comparative study of land use problems and policy solutions found in three major groups of countries. Related to U.N. Habitat Declarations.

History and Myth. Heick, Welf H. (Editor). cloth, $16.95 Tr. These essays highlight the valuable contribution made to Canadian culture by historian A.R.M. Lower, who concerned himself with the issues of everyday Canadian life.

Improving Human Settlements. Oberlander, Peter (Editor). paper, $6.95 Tr. An excellent series of lectures, pre-Habitat '76 which illuminate worldwide issues of urbanization and the search for solutions.

Mankind's Future in the Pacific. Scagel, R.F. (Editor). paper, $6.95 Tr. Thirteen of the world's most distinguished scientists discuss the future of mankind in and around the Pacific Ocean.

Colony and Confederation. Woodcock, George (Editor). paper, $6.50 Tr. A survey of the narrative and lyric poets of Confederation and the later 19th century.

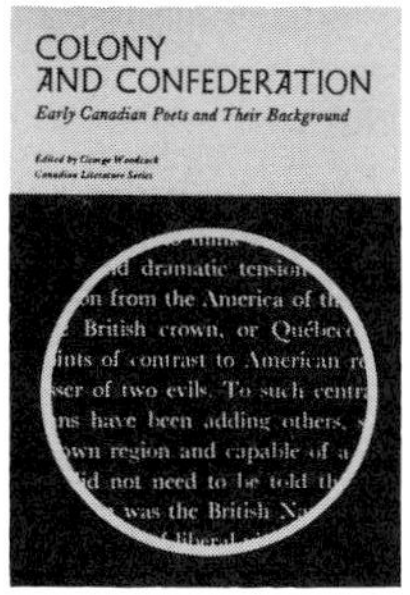

Recently Published Scholarly Titles

1979

Linguistics

Linguistic Studies of Native Canada

Edited by Eung-Do Cook and Jonathan Kaye

North American Indian languages have been in the past a somewhat obscure area of linguistic study. This volume makes an important contribution to its field by promoting increased attention to the native languages.

The fifteen articles which constitute this volume represent almost all major language families of Native Canada. It includes material on the Dakota, Sarcee, Bella Coola, Micmac, Blackfoot, Ojibwa, Cree, Nitinat, Gitksan, and Inuit languages. Each article is intensive, taking a detailed look at a particular aspect of the structure of the language, rather than extensive.

By and large the studies are based on fieldwork by the authors. All types of linguistic analysis are covered — phonetics, phonology, morphology, syntax, and semantics — and the aspects described have never been treated in print before. The application of transformational-generative theory makes the interpretations of both new and old materials in these papers highly innovative. The essays will be valuable not only to those interested in the native languages of Canada, but also to those concerned with the languages of the Western Hemisphere, with linguistic theory, and with the notation of universals.

EUNG-DO COOK is professor of linguistics at the University of Calgary.

JONATHAN KAYE is professor of linguistics at the Université du Québec à Montréal.

285 pages, 16.5 x 24 cm, 1978
ISBN 0-7748-0066-6
paper, **$24.00**
FOR SALE IN NORTH AMERICA ONLY

Raw Materials and Pacific Economic Integration

Edited by John Crawford and Saburo Okita with the assistance of Peter Drysdale and Kiyoshi Kojima

Australian-Japanese trade ranks among the world's seven largest trade flows. This trade was a crucial element in facilitating Japan's heavy industrialization and remains the cornerstone of international economic relations upon which Japan's industrial economy is founded.

This book examines in detail the gains from this economic interdependence, the problems of instability in trade, and politically sensitive areas of adjustment. Its focus is on trade in raw materials, the basis not only of Australian-Japanese trade, but also of economic relations within the Western Pacific sphere.

In addition, this comprehensive, quantitative analysis explores the nature of the interdependence existing between Japan, Australia, and the developing countries in the Western Pacific and Southeast Asia. The editors also review American attitudes and policies towards the region.

JOHN CRAWFORD is special adviser to the International Bank for Reconstruction and Development.

SABURO OKITA is chairman of the Overseas Economic Cooperation Fund.

336 pages, 14 x 22 cm, 1978
graphs and tables
ISBN 0-7748-0099-2
cloth, **$30.00**
FOR SALE IN NORTH AMERICA ONLY

Macro-Economic Policy in Australia: 1972-1976

Ainsley Jolley

The years 1972 to 1976 were an interesting and, in many respects, unique period in Australia's economic history. Inflation reached serious proportions, unemployment rose to the highest level since the Great Depression, and instability characterized commodity and financial markets. This period was marked by extraordinary fluctuations in government economic policy.

The opening chapters review how Australian economic policy is formed — the powers of the Commonwealth Government and the role of economic advisers. The Labor government's macro-economic policies are analysed for the period 1972 to 1975.

The book's main thrust is on forces determining Liberal party attitudes. Ainsley Jolley, a full-time economic adviser to the party at that time, traces the importance of such factors as election commitments, parliamentary leadership, public service advice, and the views of businessmen and academics as influences on the conduct of economic policy in 1976.

This book offers the first analysis of Australia's macro-economic policy and provides the raw material for an historical evaluation of its effectiveness.

AINSLEY JOLLEY is with the department of economics at the University of Melbourne. He is also a member of the Commonwealth Government's economic panel.

320 pages, 14 x 22 cm, 1978
tables
ISBN 0-7748-0098-4
cloth, **$32.00**
FOR SALE IN NORTH AMERICA ONLY

Western Canada Since 1870

A Select Bibliography and Guide

Alan F.J. Artibise

"The compiler's object, fully met, is to provide students, teachers, researchers, librarians, and the public with a guide to significant literature dealing with Western Canada since 1870....superb indexes are the keys to the several doors of this book. Highly recommended for academic libraries."

CHOICE

Western Canada Since 1870 is designed as a ready, reliable resource guide for students, teachers, researchers, and librarians. It is the first to cover in one volume the most significant literature in the social sciences and humanities on all four western provinces up to 1977.

Not restricted to history, it includes material that will make it a valuable research tool in the fields of geography, sociology, economics, political science, and anthropology, and native, regional, urban, and environmental studies.

The 3800 entries include periodical articles and theses as well as books and bibliographies, organized into 63 categories by topic and by province. In addition, the *General* categories include material which, while not restricted to the West, is valuable to students of the region.

The book contains author, select subject, and organization, institution, and serial indexes.

ALAN F.J. ARTIBISE is associate professor of history at the University of Victoria.

312 pages, 15 x 23 cm, 1978
illustrated
ISBN 0-7748-0090-9
cloth, **$17.50**
ISBN 0-7748-0091-7
paper, **$6.95 Tr.**

Pacifism in Japan

The Christian and Socialist Tradition

Edited by Nobuya Bamba and John F. Howes

The subject of pacifism is still relatively unexplored by scholars. It is therefore little known that one of the largest literatures on the subject, outside of Europe and North America where pacifism first developed, is in Japan.

The Sino-Japanese War of 1894 marked the end of more than two hundred years of peace in Japan. After the ensuing 1904 war with Russia, pacifism was an accepted Christian and socialist doctrine until Japan's invasion of Manchuria in 1931 made such views politically unpopular. This situation continued until 1945 when the Allied administration actively fostered anti-war sentiment as part of its programme. In 1947 a non-aggression clause was included in Japan's constitution, and pacifist activity since that time has been concerned with preserving this clause.

Pacifism in Japan: The Christian and Socialist Tradition contains eight essays which deal, among others, with such outstanding figures as Uchimura Kanzō and Kagawa Toyohiko. It is an important contribution to the understanding of the pacifist tradition and shows its development since the end of the nineteenth century. It will be of interest not only to the specialist in Japanese studies, but also to those concerned with war and peace in the modern world.

NOBUYA BAMBA specializes in Japanese diplomatic and intellectual history; formerly at McGill University, he is now a professor at Tsuda College, Tokyo.

JOHN HOWES teaches Japanese history at the University of British Columbia.

314 pages, 15 x 23 cm, 1978
ISBN 0-7748-0088-7
paper, **$9.50**
NOT FOR SALE IN JAPAN

Sales Information

Dealer's Discount Policy will be supplied on request. **"Tr"** following the price indicates a trade discount title, and no mark following the price indicates a short discount.

Dealer's Returns Policy: 100% credit allowed on books returned in mint condition at the dealer's expense. Returns are not accepted before 90 days or after one year. Permission is to be requested and received and the invoice number quoted.

Terms: net 30 days.

Publisher's ISBN prefix is 0-7748-

No handling and shipping charges on pre-paid orders.

World Rights: Unless otherwise indicated the University of British Columbia Press has world rights on all titles included in this booklist.

Prices and specifications are subject to change without notice.

Examination Copy Policy

Examination copies of books for classroom use are available on request and are accompanied by an invoice which allows up to 60 days for consideration of the text. The invoice will be cancelled if the book is adopted for the course and ten or more copies are ordered. The name of the course and estimated enrolment should be supplied. Examination copies should be requested on institution letterhead.

Sales Representation

INTERNATIONAL

United Kingdom, Northern Europe, and Scandinavia

DON FULLERTON
SPECIALISED BOOK MARKETING
22 Newhaven Road
Evington, Leicester
England LE5 6JG
Telephone: 0533-415238

Australasia

AUSTRALIA AND NEW ZEALAND BOOK COMPANY PTY. LTD.
23 Cross Street, Brookvale
N.S.W. 2100, Australia
Telephone: 938-2244

South Asia – (Japan, Korea, Hong Kong, Taiwan, Singapore, Thailand, Malaysia, the Philippines, India, and Indonesia)

PAN KOREA BOOK CORPORATION
P.O. Box no. 101
Kwangwhamum
Seoul, Korea
Telephone: 72-7369, 74-4759, 7572

CANADA

British Columbia

CLAUDETTE MARTIN
316 West 21st Street
North Vancouver
British Columbia, Canada V7M 1Z5
Telephone: (604) 980-7990

Toronto and S. Ontario

EVAN ROPER
Belford Books
78 Broadview, Toronto
Ontario, Canada M5M 3H3
Telephone: (416) 463-5981

Educational Representative
Ontario and Quebec

BRIAN DONAT
Scholarly Book Services
R.R. 4
Acton, Ontario L7J 2M1
Telephone: (519) 853-0273
(416) 877-1485

Orders from All Other Countries

are to be directed to:

THE UNIVERSITY OF BRITISH COLUMBIA PRESS
2075 Wesbrook Mall
Vancouver, British Columbia
Canada V6T 1W5
Telephone: (604) 228-3259

University of British Columbia Press books and journals may be obtained through booksellers and jobbers. Should you have any difficulty in obtaining our publications you may order directly from the University of British Columbia Press.

Order Form

Return To:

THE UNIVERSITY OF BRITISH COLUMBIA PRESS
2075 Wesbrook Mall
Vancouver, British Columbia
Canada V6T 1W5

Please send me:

Atlas of British Columbia
Quantity ______ Price $45.00 Tr.
The Canadian Yearbook, Vol. XVI
Quantity ______ Price $22.00
Canadians on the Nile, 1882-1898
Quantity ______ Price $14.95 Tr.
The Challenge of Squatter Settlements
Quantity ______ Price $4.95 Tr.
Colony and Confederation
Quantity ______ Price $6.50 Tr.
Contact and Conflict
Quantity ______ Price $18.00
Quantity ______ Price $8.95
Early Indian Village Churches
Quantity ______ Price $29.00
Quantity ______ Price $15.00 Tr.
Exploring Vancouver 2
Quantity ______ Price $7.95 Tr.
For Most Conspicuous Bravery
Quantity ______ Price $15.95 Tr.
God's Galloping Girl
Quantity ______ Price $22.00 Tr.
Habitat and Land
Quantity ______ Price $4.95 Tr.
Harsh and Lovely Land
Quantity ______ Price $18.50 Tr.
History and Myth
Quantity ______ Price $16.95 Tr.
Improving Human Settlements
Quantity ______ Price $6.95 Tr.
Indian Masterpieces
Quantity ______ Price $4.95 Tr.
The Japanese Working Man
Quantity ______ Price about $16.00
Linguistic Studies of Native Canada
Quantity ______ Price $24.00
Macro-Economic Policy
Quantity ______ Price $32.00
Malaysian Mosaic
Quantity ______ Price T.B.A.
Mankind's Future in the Pacific
Quantity ______ Price $6.95 Tr.
Northwest Coast Indian Artifacts
Quantity ______ Price $4.95 Tr.
Ocean of Destiny
Quantity ______ Price $16.50 Tr.
One Century Later
Quantity ______ Price $16.50 Tr.
Quantity ______ Price $6.95 Tr.
Pacificism in Japan
Quantity ______ Price $9.50
A Pioneer Gentlewoman
Quantity ______ Price $18.95 Tr.
Quantity ______ Price $7.95 Tr.
Planning and Building
Quantity ______ Price $4.95 Tr.
A Political Art
Quantity ______ Price $15.00 Tr.
The Reminiscences of Helmcken
Quantity ______ Price $18.95 Tr.
The Royal Navy and the Northwest Coast
Quantity ______ Price $14.00 Tr.
Tommy Brayshaw
Quantity ______ Price $32.00 Tr.
Transport in Western Siberia
Quantity ______ Price $22.00
Visitors Who Never Left
Quantity ______ Price $5.95 Tr.
Western Canada Since 1870
Quantity ______ Price $17.50
Quantity ______ Price $6.95 Tr.
The Wind Commands
Quantity ______ Price $29.00 Tr.

☐ Payment enclosed (No handling and shipping charges on pre-paid orders).

☐ Invoice me

NAME: ______________________

ADDRESS: ______________________

____

____

☐ Put me on your mailing list

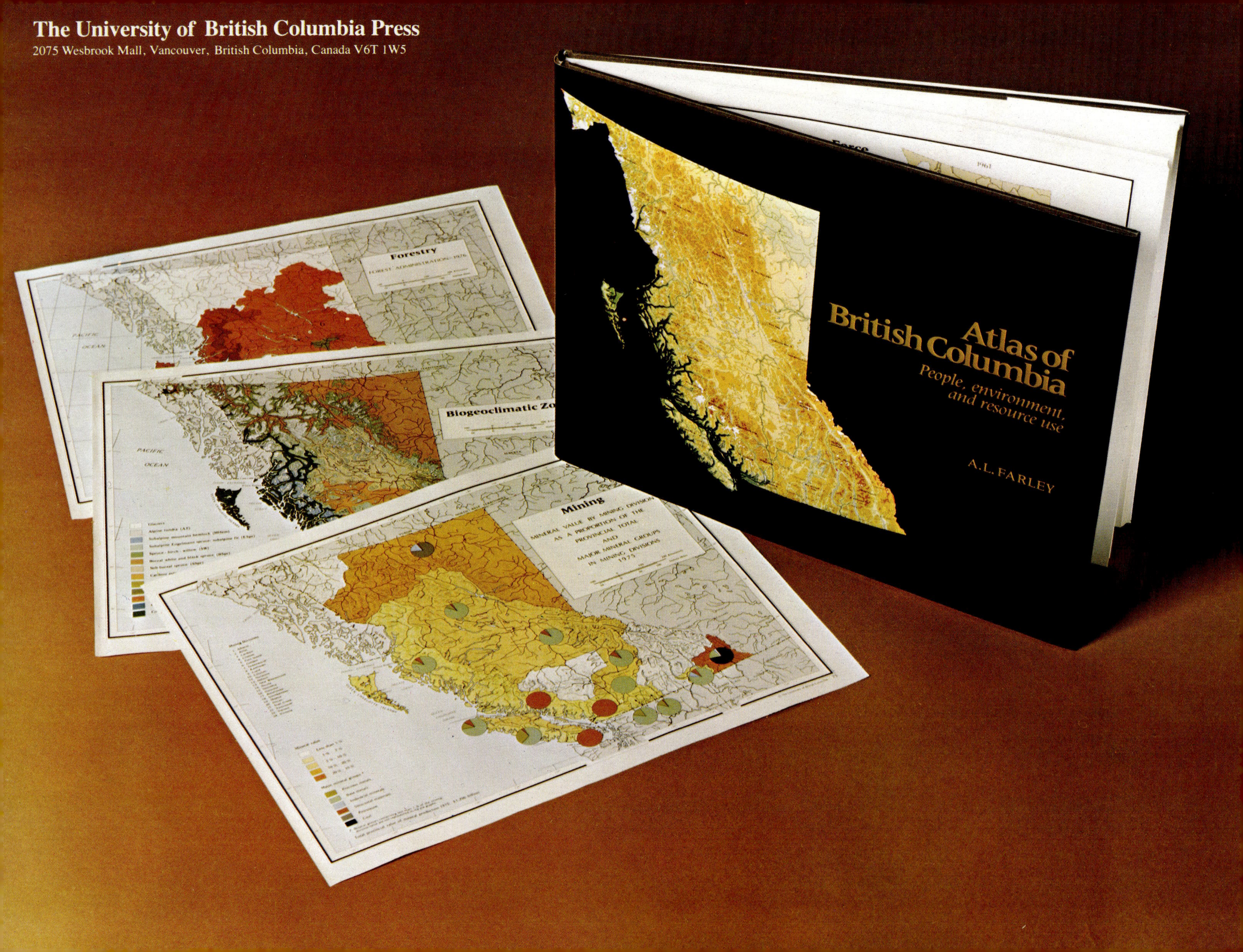
The University of British Columbia Press
2075 Wesbrook Mall, Vancouver, British Columbia, Canada V6T 1W5
Forestry
Biogeoclimatic Zo
Mining
MINERAL VALUE BY MINING DIVISION
AS A PROPORTION OF THE
PROVINCIAL TOTAL
AND
MAJOR MINERAL GROUPS
IN MINING DIVISIONS
1975
PACIFIC OCEAN
Atlas of
British Columbia
People, environment,
and resource use
A.L. FARLEY

Atlas of British Columbia
People, environment, and resource use

AUTHOR'S BIOGRAPHY

The University of British Columbia Press
2075 WESBROOK MALL, VANCOUVER, BRITISH COLUMBIA, CANADA, V6T 1W5
TELEPHONE (604) 228-3259 / CABLE: UNIPRESS VANCOUVER

Atlas of British Columbia
People, environment, and resource use

AUTHOR'S BIOGRAPHY Dr. A.L. (Bert) Farley

Dr. Farley is a native of Victoria, B.C. He was born in 1920 and after finishing school and teacher training there, he entered the Canadian Navy during the Second World War, serving on Corvettes in the North Atlantic.

Following his demobilization, he returned to B.C. and took up his studies at the University of British Columbia, where he received his B.A. and M.A. in geography. He then undertook doctoral studies at the University of Wisconsin in the School of Academic Cartography.

Upon completion of his Ph.D., Dr. Farley accepted a position with the B.C. provincial government, working for five years as a geographer in the Surveys and Mapping Branch.

In 1958 he became a member of the Geography faculty at UBC where he teaches today. His courses specialize in the geography of B.C. and the Pacific Northwest, Cartography, and Remote Sensing (the analysis of aerial photographs).

As a member of the B.C. Natural Resources Conference, Dr. Farley was given responsibility as the cartographic editor for the 1956 Atlas which resulted from that Conference.

He serves on the National Cartographic Commission, is an associate editor for Canadian Survey, and he sits on the Exectuive of the Canadian Cartographic Association. He is a member of the American Geography Society and the Canadian Association of Geographers.

The University of British Columbia Press
2075 WESBROOK MALL, VANCOUVER, BRITISH COLUMBIA, CANADA, V6T 1W5
TELEPHONE (604) 228-3259 / CABLE: UNIPRESS VANCOUVER

Atlas of British Columbia

People, environment, and resource use

ADVANCE REVIEW

The following review has been kindly contributed by John Schreiner of the Financial Post's "Western Business." Should any editor wish to reprint it in full, or excerpted, as required, they are at liberty to do so, with credit to Mr. Schreiner.

PLEASE DO NOT FEATURE A REVIEW OF THIS BOOK PRIOR TO THE PUBLICATION DATE - JUNE 25, 1979

The University of British Columbia Press
2075 WESBROOK MALL, VANCOUVER, BRITISH COLUMBIA, CANADA, V6T 1W5
TELEPHONE (604) 228-3259 / CABLE: UNIPRESS VANCOUVER

The Financial Post

Canada's national journal of business, investment and public affairs

ATLAS OF BRITISH COLUMBIA: People, Environment, and Resource Use

By A.L. Farley: University of British Columbia Press

The following review has been kindly contributed by John Schreiner, distinguished Editor of the Financial Post's "Western Business." It may be reprinted in full, or excerpted, as required, with credit to Mr. Schreiner.

By repute, Texans boast about the bigness of their state. Who knows what they would do if they had the stunning scale of British Columbia, as revealed by this atlas in words and diagrams.

Just how big is the Fraser River? "The Fraser River drainage basin occupies about one-quarter of the total area of the province," writes author Farley, a geographer at the University of British Columbia.

Just what sort of earthquakes does British Columbia have? It seems they are shallow, with epicentres on average about 16 kilometers below the surface. But the shallow ones are more likely to cause localized surface damage than the deep ones, even if the latter are widely felt. A quick and comforting glance at one map in the atlas shows that the major quake belt parallels the coast about 200 miles offshore.

The most important newsprint mill in the province? It is the one at Powell River. A photograph in the atlas shows the mill under construction in 1911. The facilities there today are operated by MacMillan Bloedel and continue to be expanded.

- more -

These are three examples of the information nuggets crammed in a book which is handsome and readable as well as scholarly. Farley has been working on it since 1975, as a replacement for the 1956 British Columbia Atlas of Resources, on which he was a cartographer.

The atlas sets some of the social scene in its historic perspective, with good maps showing how British Columbia was explored, where the various Indian groups lived and where today's reserves are. Recent migration trends also are shown, dramatizing how the province has attracted people and where they've come from.

The 115 maps -- nearly all of which use colour to illustrate the information -- range across the high points of the provincial economy, from energy and mining to transportation and agriculture. In short, the atlas is a solid contribution to understanding British Columbia.

- end -

PRE-PUBLICATION COMMENT FROM THE
INDUSTRIAL AND PROFESSIONAL COMMUNTTIES
OF B.C.

The University of British Columbia Press
2075 WESBROOK MALL, VANCOUVER, BRITISH COLUMBIA, CANADA, V6T 1W5
TELEPHONE (604) 228-3259 / CABLE: UNIPRESS VANCOUVER

BRITISH COLUMBIA HYDRO AND POWER AUTHORITY

ROBERT W. BONNER, Q.C.
CHAIRMAN

970 BURRARD STREET
VANCOUVER, B. C.
V6Z 1Y3

The new Atlas of British Columbia should be required reading in every household in the province because without some general grasp of the detail which it records it is almost impossible to understand the provincial economy or the nature of the work and opportunities upon which the lives of our people are based.

British Columbia, like most parts of Canada, has a domestic resource-based economy. The Atlas of British Columbia records in most useful detail the variety and richness of the resources which we have been so fortunate to inherit in this province.

Council of
Forest Industries
of British Columbia

D. A. S. LANSKAIL
President

March 13, 1979

Getting to know more about the part of the world we inhabit is a continuing project for many. For some it is an interesting hobby but for others it is a necessary activity if we are to reconcile the varying, and sometimes competing social and economic interests that determine our lifestyle. The new Atlas of B.C. is an important resource document to aid in gaining that knowledge.

The sections on the history, growth and relative economic importance to B.C. of the forest industry will be of particular value to students. Graphically illustrated, too, are the formidable problems of transportation confronting the resources industries of this province.

The new Atlas of B.C. is a useful educational tool and a very interesting historical record of the province and its development.

D. A. S. Lanskail
President and
Chief Executive Officer

1500/1055 West Hastings St., Vancouver, Canada V6E 2H1 Tel. (604) 684-0211, Cable: CANWOOD Vancouver

THE MINING ASSOCIATION OF BRITISH COLUMBIA
104 - 1075 Melville Street, Vancouver, B.C., Canada V6E 2W4 Phone (604) 681-3374 Telex 04-507784

The new Atlas of British Columbia is a very valuable and excellently presented picture of British Columbia, its peoples and its resources. The section on mining very clearly and graphically portrays the present status as well as the history of the mineral industry.

Mining, which ranks second only to forestry in its contribution to the basic wealth of the province, has been essential to British Columbia since its earliest days. Despite the non-renewable nature of the resource and the cyclical pattern of the international markets on which it depends, the industry has remained a major component of the Provincial economy. The effect of these factors, the changes and growth in the industry and the future for mining in British Columbia are concisely explained in the text, charts and maps.

J. H. Parliament
President
Mining Association of British Columbia

MEMBER
THE FISHERIES COUNCIL OF CANADA

FISHERIES ASSOCIATION OF BRITISH COLUMBIA

Room 400 - 100 W. Pender Street, Vancouver, B.C. V6B 1R8 Tel. (604) 684-6454

CANNED SALMON ASSOCIATION OF BRITISH COLUMBIA

March 28, 1979

Students, teachers and other persons interested in getting to know more about our province will find the "Atlas of British Columbia, People, Environment and Resource Use" an invaluable aid.

The fishing industry is making a greater and greater contribution to British Columbia's natural resource oriented economy. The Atlas graphically portrays many characteristics of the fishing industry in a clear and concise manner.

THE FISHERIES ASSOCIATION OF BRITISH COLUMBIA

J. N. Spitz,
President

JNS;bjs

846 Broughton street,
Victoria, B.C. V8W 1E4
(604) 383-7171

BRITISH COLUMBIA FEDERATION OF AGRICULTURE

It is with a great deal of enthusiasm that the agricultural industry in British Columbia observes the publication of a new atlas of British Columbia.

Like all of the sectors which make up our modern economy, the farming community is acutely aware of the complexity of current resource issues and of the subsequent importance of well researched and accessible information in assisting us to embark upon the path that will best reflect our needs and values. On behalf of the B.C. Federation of Agriculture I would like to congratulate Professor Farley and all those who participated in the production of the atlas for the substantial contribution which their work will make to this process. The information assembled in this book has an important role to play in enabling British Columbians to gain a better understanding of the varied and rich resources which constitute both our past and our future.

Pat Hibbert
President
B.C. Federation of Agriculture

British Columbia
teachers' federation

105 - 2235 BURRARD STREET,
VANCOUVER, BRITISH COLUMBIA,
V6J 3H9
TELEPHONE (604) 731-8121
General Secretary
R.M. Buzza

The *British Columbia Atlas of Resources*, 1956, became a standard reference used in schools, particularly secondary schools, of our province. The *Atlas of British Columbia: People, Environment and Resource Use*, 1979, should serve the same important purpose. It provides a comprehensive progress report on migration and environmental and resource use trends during the 20-year period, 1956 to 1976.

Students and teachers will welcome this new atlas.

P. T. Brady

Patrick T. Brady,
President.

PTB:am

association of
PROFESSIONAL ECONOMISTS
OF BRITISH COLUMBIA

The Atlas of B.C. is an excellent reference as a source of information about B.C., its people and its resources.

The population statistics together with the excellent work on forestry and mining - the two most important industries in the province - will prove invaluable to businessmen and students alike who are searching for information about B.C. and its economy.

Gordon H. MacDougall
President

THE ASSOCIATION OF
PROFESSIONAL ENGINEERS
OF THE PROVINCE OF BRITISH COLUMBIA

2210 WEST 12TH AVENUE, VANCOUVER, B.C. V6K 2N6 PHONE: 736-9808

March 16, 1979

Our environment has always been of concern to Professional Engineers. Today's form came to us as a result of yesterday's combination of men, materials and attitudes; tomorrow's shape will be the result of today's decisions.

As a group, Professional Engineers are most aware of the need to understand and use wisely the resources of nature and man. I would therefore encourage the public to use this excellent atlas in order to enhance not only their knowledge but also their individual skills in contributing to the future of British Columbia.

R. G. Fraser, P.Eng.,
President.

Atlas of British Columbia
People, environment, and resource use

FILLER FACTS & ODDITIES FROM THE ATLAS

The University of British Columbia Press
2075 WESBROOK MALL, VANCOUVER, BRITISH COLUMBIA, CANADA, V6T 1W5
TELEPHONE (604) 228-3259 / CABLE: UNIPRESS VANCOUVER

Atlas of British Columbia
People, environment, and resource use

DO YOU <u>REALLY</u> KNOW BRITISH COLUMBIA??

Filler facts and oddities from the Atlas of B.C.

1. British Columbia accounts for 9.5% of the total area of Canada.

2. B.C. shares a 45-mile boundary with the state of Idaho, and a 960-mile border with Alberta. (72 km and 1545 km)

3. Both Edmonton and Prince George receive the same number of telephone calls from Dawson Creek.

4. There are as many people living in Vancouver west of Granville Street as in the entire Kootenay-Boundary area.

5. Only 3% of B.C.'s land area is capable of growing crops.

6. The B.C. record for a steelhead trout is 17 kilograms (37.4 lbs.).

7. The largest lake in B.C. is man-made and three times larger than the greatest natural lake entirely within B.C. Williston Reservoir is 1750 sq. km (675 sq. mi.), while Babine Lake is 497 sq. km (192 sq. mi.) including islands.

8. Besides Canadian National and Canadian Pacific Railways, there are five others operating in the province - BC Railway, Northern Alberta Rly, Burlington Northern Rly, White Pass and Yukon Rly, and two others operated by B.C. Hydro and the B.C. Harbours Board.

9. Della Falls on Vancouver Island, classified as the highest waterfall in B.C. at 440 m (1443 ft.), is actually a cascade, or series of smaller falls.

10. The combined areas of provincial and national parks in B.C. provide for 2 hectares (nearly 5 acres) for every man, woman and child in the Province.

- more -

Oddities: cntd.

11. Exports from B.C. account for 17% of Canada's total.

12. Although Mount Waddington, at 4016 m (13,175 ft.) is the highest peak in B.C. it is not the highest elevation. That honour goes to Fairweather Mountain at 4663 m (15,298 ft.), whose peak is located outside B.C.

13. There are approximately 350 landing fields and 100 seaplane bases in B.C.

14. There are about 6500 coastal islands in B.C. waters, and that does not include "islets."

15. Kimberley enjoys the highest city elevation, at 1115 m (3,658 ft.).

16. In 1911 Strathcona Park was established as the first Provincial Park in B.C.

17. Dewdney began his famous trail from Hope to Similkameen in 1860.

18. B.C.'s mainland coast measures about 12,150 km (7551 miles), not counting the shores of islands.

19. The deepest sounding in coastal waters is in the Finlayson Channel, where a depth of 764 m (2508 ft.) has been recorded.

20. After Vancouver Island, of course, the largest island in B.C. waters is Graham Island in the Queen Charlottes, at 6436 sq km (2484 sq. miles).

21. B.C.'s longest river, the Fraser, at about 1370 km (851 miles), has a drainage area of approximately 230,000 sq. km.

22. B.C. provides 4.5% of the world's supply of pulp.

23. The longest distance, north to south, in the province is about 1300 km (808 miles), and the longest from east to west is 1062 km (660 miles).

- more -

Oddities: cntd.

24. The shortest distance, north to south, is 24 km (14.9 miles), and the shortest east to west is about 625 km (388 miles).

25. The most southerly point in the province is Rosedale Rock off southern Vancouver Island, at latitude 48°17'34" N.

Oxford English Dictionary Definition of a Map

"A representation of the earth's surface or a part of it, its physical features, etc., or of the heavens, delineated on a flat surface of paper or other material, each point in the drawing corresponding to a geographical or celestial position according to a definite scale or projection."

- end -

Atlas of British Columbia

People, environment, and resource use

UBC PRESS

The University of British Columbia Press 2075 WESBROOK MALL, VANCOUVER, BRITISH COLUMBIA, CANADA, V6T 1W5
TELEPHONE (604) 228-3259 / CABLE: UNIPRESS VANCOUVER

UNIVERSITY OF BRITISH COLUMBIA PRESS
2075 WESBROOK MALL
VANCOUVER, BRITISH COLUMBIA
CANADA V6T 1W5

TELEPHONE (604)228-3259 CABLE: UNIPRESS VANCOUVER

BACKGROUND

THE UNIVERSITY PRESS

The University of British Columbia Press was established in 1971 as an expansion of the former University Publications Centre. It was formed in response to a growing need for an outlet for much of the research being done on the west coast of Canada, where no other university press exists. The Press serves the academic community by publishing important works of scholarship in three specialized areas -- Asia and the Pacific, International Law, and Canadian Studies. It also serves the community of which it is a part by publishing books that help to preserve regional heritage and culture. UBC Press books include works on native Indian people, west coast naval history, pioneer memoirs, and resource policy. The Atlas of British Columbia is the latest work in this regional publishing programme. The Press published four titles in its first year of operation. Now, eight years later, it has more than seventy-five titles to its credit and is publishing nineteen to twenty books annually. UBC Press sells its books internationally. In 1978 alone, 24,000 copies of its books were sold in forty-nine countries.

Atlas of British Columbia
People, environment, and resource use

BASIC NEWS RELEASE

The University of British Columbia Press
2075 WESBROOK MALL, VANCOUVER, BRITISH COLUMBIA, CANADA, V6T 1W5
TELEPHONE (604) 228-3259 / CABLE: UNIPRESS VANCOUVER

UNIVERSITY OF BRITISH COLUMBIA PRESS

2075 WESBROOK MALL
VANCOUVER, BRITISH COLUMBIA
CANADA V6T 1W5

TELEPHONE (604) 228-3259

CABLE: UNIPRESS VANCOUVER

REVIEW COPY

Title: ATLAS OF BRITISH COLUMBIA

Sub-Title: People, Environment, and Resource Use

Author(s): A.L. Farley

Editor(s):

Pages: 144 Size: 28 x 32.5 cm (11 x 14 in.)

Illustrations: 33 colour, 41 black and white photographs
115 maps, 11 tables, 16 diagrams

ISBN (cloth): 0-7748-0092-5 (paper):

Other:

Price (cloth): $45.00 Tr. (paper):

PUBLICATION DATE: June 25, 1979

Please do not feature a review of this book prior to publication.

Queries: call the Marketing and Promotion Department at (604) 228-3259

This book is sent to you with the compliments of the University of British Columbia Press. If you review this book we would appreciate receiving two copies of your printed review. Thank you.

News Release

12 June 1979

FOR IMMEDIATE RELEASE

Director of the University of British Columbia Press, Mr. Anthony Blicq, today announced completion of a four-year project: publication of a new ATLAS OF BRITISH COLUMBIA.

Describing it as a "major event in the publishing history of this province," Blicq explained that the new ATLAS will replace the 1956 British Columbia Atlas of Resources. It will serve as a basic informational tool for industrial, government and educational use well into the 1980's.

First run printing is of twenty-five thousand (25,000) copies, ten thousand (10,000) of which are destined for the schools of B.C. Sales of the $45 book are expected to be brisk, according to the Press' marketing manager, Ann Hockey.

Given the changes and developments in the B.C. picture over the past twenty years, provincial government interest in an updated ATLAS has been constant and encouraging since the project began in 1975. In his "Foreword" Premier Bill Bennett states, "it was with pleasure that the provincial government joined with industry and the university to help bring the Atlas to completion...this new Atlas will be of lasting value for the people of the province."

Work was begun in 1975. Under direction from author Dr. A.L. (Bert) Farley of UBC's Department of Geography, a group of students helped to prepare the maps. Assistance from the provincial Youth Employment Program and the federal Young Canada Works project enabled these students to gain direct practical experience within the university on a large-scale cartographic undertaking.

- more -

The University of British Columbia Press
2075 WESBROOK MALL, VANCOUVER, BRITISH COLUMBIA, CANADA, V6T 1W5
TELEPHONE (604) 228-3259 / CABLE: UNIPRESS VANCOUVER

The ATLAS employs the latest cartographic techniques in its 113 full-colour maps, uses the metric system, and its contents are in part the result of public needs and requests made through libraries and through a testing programme with the general public carried out during UBC's OPEN HOUSE in 1976.

Commenting on the significance of this publication, Blicq observed that "in terms of size and costs, this is the single most ambitious publication from the University Press since its foundation in 1971.

"It is one of our principal objectives to provide books about B.C. for British Columbians, and this ATLAS represents a fine achievement of the practical operation of that policy. We expect it to serve the community for many years to come, and I take great pride in announcing its completion."

- end -

For further information: Mr. Anthony Blicq (604) 228-3807
Miss Ann Hockey (604) 228-3259
Dr. A.L. Farley Work: (604) 228-3534
Home: (604) 224-0217

